冥王星

以及其他矮行星

PLUTO

and the Dwarf Planets

（英国）埃伦·劳伦斯／著　　张　骁／译

U0240963

江苏凤凰美术出版社

著作权合同登记图字：10-2022-144

图书在版编目（CIP）数据

冥王星：以及其他矮行星 /（英）埃伦·劳伦斯著；

张骁译 . -- 南京：江苏凤凰美术出版社，2025. 1.

（环游太空）. -- ISBN 978-7-5741-2027-3

Ⅰ . P185.6-49

中国国家版本馆 CIP 数据核字第 2024HD2064 号

策　　　　划　朱　婧

责 任 编 辑　高　静　奚　鑫

责 任 校 对　王　璇

责任设计编辑　樊旭颖

责 任 监 印　生　嫄

英 文 朗 读　C.A.Scully

项 目 协 助　邵楚楚　乔一文雯

丛 书 名　环游太空

书　　　名　冥王星：以及其他矮行星

著　　　者　（英国）埃伦·劳伦斯

译　　　者　张　骁

出 版 发 行　江苏凤凰美术出版社（南京市湖南路 1 号　邮编：210009）

印　　　刷　南京新世纪联盟印务有限公司

开　　　本　710 mm×1000 mm　1/16

总 印 张　18

版　　　次　2025 年 1 月第 1 版

印　　　次　2025 年 1 月第 1 次印刷

标 准 书 号　ISBN 978-7-5741-2027-3

总 定 价　198.00 元（全 12 册）

版权所有　侵权必究

营销部电话：025-68155675　营销部地址：南京市湖南路 1 号

江苏凤凰美术出版社图书凡印装错误可向承印厂调换

目录 Contents

书中加粗的词语见词汇表解释。

Words shown in **bold** in the text are explained in the glossary.

欢迎来到冥王星
Welcome to Pluto

想象一下，你正在飞往一个离地球好几十亿千米的星球。

Imagine flying to a world that is billions of kilometers from Earth.

无论从哪个方向看，这片土地上都覆盖着冰层和岩石。

In every direction there is rocky, icy land.

这个渺小又遥远的世界比地球上最冷的地方还要冷得多。

On this tiny, faraway world, it is much colder than the coldest place on Earth.

欢迎来到矮行星——冥王星！

Welcome to the **dwarf planet**, Pluto!

冥王星 Pluto

地球 Earth

火星 Mars

矮行星是一种体积较小的圆形天体。它比一些大行星，比如地球和火星要小得多。

A dwarf planet is a small, round space object. Dwarf planets are much smaller than big **planets**, such as Earth and Mars.

我们的家园——地球只有一个卫星，但是小小的冥王星有5个！冥王星最大的卫星叫作卡戎（冥卫一）。

Our home planet Earth has one **moon**, but tiny Pluto has five! Pluto's largest moon is called Charon (KARE-uhn).

卡戎 Charon

冥王星表面
The surface of Pluto

这些冥王星表面和卡戎的照片是由"新视野号"太空探测器于2015年飞越冥王星时拍摄的。

These photos of Pluto's surface and Charon were taken by the *New Horizons* space **probe** as it flew by Pluto in 2015.

太阳系 The Solar System

冥王星围绕着太阳做一个巨大的圆周运动。

Pluto is moving in a huge circle around the Sun.

其他矮行星也绕着太阳公转。

Other dwarf planets are **orbiting**, or moving, around the Sun, too.

还有八大行星也在围绕着太阳旋转。

There are also eight big planets circling the Sun.

它们分别是水星、金星、我们的母星地球、火星、木星、土星、天王星和海王星。

The big planets are called Mercury, Venus, our home planet Earth, Mars, Jupiter, Saturn, Uranus, and Neptune.

这些行星、矮行星和其他天体共同组成了"太阳系"。

Together, the planets, dwarf planets, and other space objects are called the **solar system**.

结满冰的彗星和被称为"小行星"的太空岩石也围绕着太阳公转。太阳系中的大多数小行星都集中在被称为"小行星带"的环状带中。

Icy **comets** and space rocks called **asteroids** also circle the Sun. Most of the asteroids are in a ring called the **asteroid belt**.

小行星 An asteroid

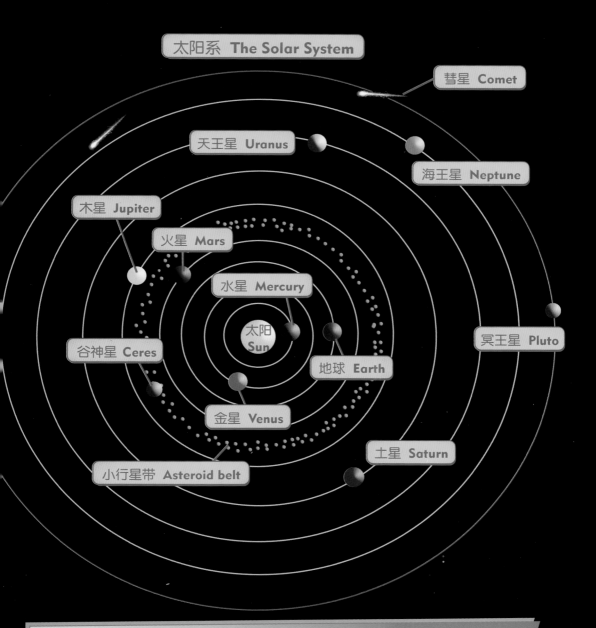

太阳系 The Solar System

彗星 Comet

天王星 Uranus

海王星 Neptune

木星 Jupiter

火星 Mars

水星 Mercury

太阳 Sun

地球 Earth

谷神星 Ceres

冥王星 Pluto

金星 Venus

小行星带 Asteroid belt

土星 Saturn

在这幅图中我们可以看到两颗矮行星——冥王星和谷神星。其他矮行星则在冥王星以外的空间中运动。

Two dwarf planets, Pluto and Ceres, can be seen in this diagram. Other dwarf planets are moving through space beyond Pluto.

遇见矮行星
Meet the Dwarf Planets

多年来，冥王星都被认为是行星。

For many years, Pluto was called a planet.

后来，科学家们开始发现其他和冥王星一样环绕太阳公转的遥远小天体。

Then, scientists began to discover other small objects like Pluto orbiting far away from the Sun.

于是他们决定把这些小天体单独分为一类，并把它们称为矮行星。

They decided to put these small objects into their own group and call them dwarf planets.

首批被划为矮行星的天体有冥王星、阋神星、谷神星、鸟神星和妊神星。

The first five space objects to be called dwarf planets were Pluto, Eris, Ceres, Makemake, and Haumea.

科学家们还在研究其他被认为可能是矮行星的天体。

Scientists are studying other space objects which they think could be dwarf planets.

还有更多矮行星有待发现。

It's likely there are many more still to be discovered.

矮行星的名字 Dwarf Planet Names

谷神星	Ceres (SIHR-eez)
阋神星	Eris (IHR-iss)
妊神星	Haumea (how-MEH-uh)
鸟神星	Makemake (MAH-kee-MAH-kee)
冥王星	Pluto (PLOO-toh)

火星 Mars

水星 Mercury

太阳 Sun

谷神星 Ceres

地球 Earth

金星 Venus

小行星带 Asteroid belt

谷神星是离太阳最近的矮行星，它在小行星带上绕着太阳公转。

Ceres is the closest dwarf planet to the Sun. It orbits the Sun in the asteroid belt.

妊神星 Haumea

冥王星 Pluto

太阳 Sun

阋神星 Eris

鸟神星 Makemake

冥王星、阋神星、鸟神星和妊神星离太阳有几十亿千米远。这张图片展示了它们围绕太阳运动的轨迹。

Pluto, Eris, Makemake, and Haumea are billions of kilometers from the Sun. This picture shows the shapes of their journeys around the Sun.

近距离观察冥王星
A Closer Look at Pluto

每绕太阳一圈，冥王星都要走过极其漫长的路程。

To orbit the Sun once, Pluto makes a super-long journey.

它需要在太空中运行近370亿千米。

It travels through space for nearly 37 billion kilometers.

一个天体围绕太阳公转一圈所需的时间被称为"一年"。

The time it takes a space object to orbit, or circle, the Sun once is called its year.

地球绕太阳公转一圈需要略多于365天，所以地球上的一年有365天。

Earth takes just over 365 days to orbit the Sun, so a year on Earth lasts 365 days.

冥王星绕太阳公转一圈大约需要248个地球年。

It takes Pluto almost 248 Earth years to orbit the Sun.

这意味着，地球上的248年才相当于冥王星上的1年！

This means a year on Pluto lasts for 248 Earth years!

当矮行星围绕太阳公转时，它也像陀螺一样自转着。

As a dwarf planet orbits the Sun, it also **rotates**, or spins, like a top.

冥王星 Pluto

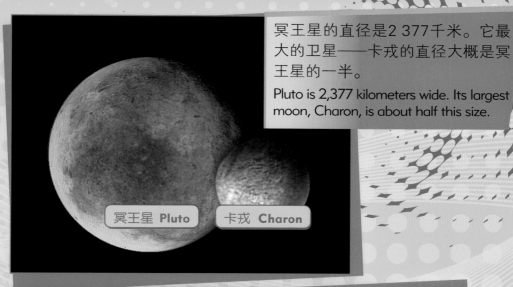

冥王星的直径是2 377千米。它最大的卫星——卡戎的直径大概是冥王星的一半。

Pluto is 2,377 kilometers wide. Its largest moon, Charon, is about half this size.

冥王星 Pluto　卡戎 Charon

太阳 The Sun

冥王星 Pluto

这张图片展示了站在冥王星上可能会看到的场景，遥远的太阳看上去就像一个闪耀的小点！

This picture shows how it might look to stand on Pluto. The faraway Sun looks like a tiny, shining dot!

遥远的阋神星
Faraway Eris

在公转轨道上运行时，阋神星离太阳的距离有时会达到地球和太阳距离的97倍。

As it orbits the Sun, Eris is some-times 97 times farther from the Sun than Earth.

这颗矮行星绕太阳一圈可能要花上560个地球年！

It takes this dwarf planet 560 Earth years to orbit the Sun once!

因为离地球太远了，阋神星的研究对于科学家们来说十分困难。

It is difficult for scientists to study Eris because it is so far from Earth.

科学家们认为阋神星一定极为寒冷，因为它太遥远了，接收不到太阳的热量。

They know it is super-cold be-cause it is so far from the Sun's heat.

他们还认为阋神星的大小应该与冥王星差不多，但是并不非常确定。

They also think it is about the same size as Pluto, but no one knows for sure.

这张图片模拟了阋神星的样子。它有一个卫星，叫作迪丝诺美亚（阋卫一）。当然啦，阋神星也可能有更多卫星等待人们发现。

This picture shows how Eris might look. It has one moon, called Dysnomia (dis-NOH-mee-uh). There may be more moons to be discovered, though.

太阳 **The Sun**

迪丝诺美亚 **Dysnomia**

阋神星 **Eris**

鸟神星和妊神星
Makemake and Haumea

妊神星是一颗蛋形的矮行星，覆盖着岩石和冰层。

它每绕太阳一圈要花283个地球年。

而鸟神星绕太阳一圈则要花上307个地球年！

冥王星、阋神星、妊神星和鸟神星围绕太阳公转的那一片区域叫"柯伊伯带"。

柯伊伯带是一个巨大的环形区域，有成千上万冰冻的天体在此聚集。

Haumea is a rocky, icy, egg-shaped dwarf planet.

It takes 283 Earth years to orbit the Sun once.

Makemake's journey around the Sun lasts for 307 Earth years!

Pluto, Eris, Haumea, and Makemake orbit the Sun in an area called the Kuiper Belt (KY-pur BELT).

The Kuiper Belt is a giant ring where thousands of icy space objects are gathered.

鸟神星 Makemake

这张照片呈现的是遥远的鸟神星。这颗矮行星上覆盖着冰层。
This photo shows faraway Makemake. This dwarf planet is covered with ice.

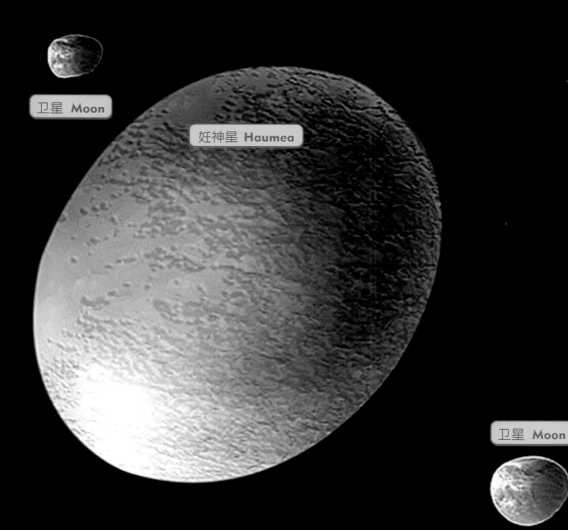

卫星 Moon

妊神星 Haumea

卫星 Moon

科学家已经在妊神星周围发现了两颗卫星，但是也可能有更多。
这张图片显示了妊神星以及它的卫星可能看上去的样子。

Scientists have found two moons around Haumea, but there may be more. This picture shows what Haumea and its moons might look like.

遇见谷神星
Meet Ceres

谷神星既是矮行星，也是小行星。

Ceres is both a dwarf planet and an asteroid.

这颗巨大的岩石天体是环绕太阳的小行星带中的一员。

This large, rocky object is circling the Sun in the asteroid belt.

谷神星比别的矮行星离太阳更近。

Ceres is much nearer to the Sun than the other dwarf planets.

它绕太阳一圈只需要花上略多于4个半地球年。

It takes just over four and a half Earth years to orbit the Sun once.

谷神星的直径大约是952千米。

Ceres is about 952 kilometers wide.

差不多和美国得克萨斯州一样宽！

That's almost as wide as Texas!

这是谷神星的照片。它是小行星带上最大的小行星。

This is a photo of Ceres. It is the biggest asteroid in the asteroid belt.

谷神星 Ceres

地球 Earth

这张图片展现出谷神星和地球的大小对比。

This picture shows the size of Ceres compared to Earth.

探测冥王星的任务
A Mission to Pluto

2006年1月，一个名为"新视野号"的空间探测器从地球发射。

In January 2006, a space probe named *New Horizons* blasted off from Earth.

这个探测器于2015年到达了冥王星，并近距离飞越这颗矮行星。

It reached Pluto in 2015 and made a **flyby** of the dwarf planet.

"新视野号"发现了冥王星表面覆盖着冰和岩石，还有蓝色的大气层。

New Horizons discovered that Pluto has ice and rock on its surface and a blue atmosphere.

它还发现冥王星的表面曾经有过液体。

It also discovered that there were once liquids on Pluto's surface.

在飞越冥王星以后，"新视野号"飞向了我们太阳系外围的柯伊伯带，并将对这个区域展开研究。

After the flyby, *New Horizons* headed for the outer edge of our solar system to study the Kuiper Belt.

"新视野号"提供的数据表明，冥王星的5颗卫星或许形成于同一时期。这意味着它们是在冥王星与其他天体相撞时形成的。这次撞击导致一些巨大的块状物从这颗矮行星上脱离，然后变成了这些卫星。

New Horizons showed that Pluto's five moons may all be the same age. This means they formed when Pluto collided with another object. The collision made huge chunks of the dwarf planet break away and become the moons.

"新视野号"用一个星期的时间拍摄了冥王星的照片并且收集了相关数据。但是这些信息却花了15个月才陆陆续续全部被传回地球！

New Horizons took pictures and collected data about Pluto for a week. It took 15 months to beam all the information back to Earth!

卡戎星
Charon

"新视野号"
New Horizons

冥王星 **Pluto**

心形冰川 **The heart glacier**

"新视野号"拍摄了这张冥王星照片。照片上这片平坦的区域被称为"冥王星之心"，是太阳系中最大的冰川平原。这座冰川平原由氮冰构成，宽度超过1 000千米。

New Horizons took this photo of Pluto. It shows a flat area called the heart which is the biggest **glacier** in the solar system. It is made of nitrogen ice and is more than 1,000 kilometers wide.

有趣的冥王星知识
Pluto Fact File

以下是一些有趣的冥王星知识：冥王星是最著名的矮行星。

Here are some key facts about Pluto, the most famous dwarf planet.

冥王星的发现
Discovery of Pluto

冥王星于1930年2月18日被天文学家克莱德·威廉·汤博发现。

Pluto was discovered on February 18, 1930, by Clyde W. Tombaugh.

冥王星是如何得名的
How Pluto got its name

冥王星是以古罗马神话中掌管冥府和来世的神的名字命名的。

Pluto is named after the Roman god of the underworld and afterlife.

行星的大小
Planet sizes

这张图显示了太阳系八大行星和冥王星的大小对比。

Here are the sizes of the solar system's eight big planets compared to Pluto.

水星 Mercury　地球 Earth　太阳 Sun　金星 Venus　火星 Mars　木星 Jupiter　土星 Saturn　天王星 Uranus　海王星 Neptune　冥王星 Pluto

冥王星的大小
Pluto's size

冥王星的直径约2 377千米
2,377 km across

冥王星自转一圈需要多长时间
How long it takes for Pluto to rotate once

大约153个地球时（大约6.5个地球天）
About 153 Earth hours (About 6.5 Earth days)

冥王星与太阳的距离
Pluto's distance from the Sun

冥王星与太阳的最短距离是4 436 756 954千米。
冥王星与太阳的最远距离是7 376 124 302千米。

The closest Pluto gets to the Sun is 4,436,756,954 km.
The farthest Pluto gets from the Sun is 7,376,124,302 km.

冥王星围绕太阳公转的平均速度
Average speed at which Pluto orbits the Sun

每小时17 100千米

17,100 km/h

冥王星的卫星
Pluto's moons

冥王星至少有5颗卫星。可能还有更多的有待发现。

Pluto has at least five moons. There are possibly more to be discovered.

冥王星绕太阳轨道的长度
Length of Pluto's orbit around the Sun

36 529 978 039千米
36,529,978,039 km

冥王星 Pluto

太阳 Sun

冥王星轨道 Pluto's orbit

冥王星上的一年
Length of a year on Pluto

略多于90 553个地球天（大约248个地球年）
Over 90,553 Earth days (Nearly 248 Earth years)

冥王星上的温度
 ## Temperature on Pluto

零下228摄氏度
-228℃

动动手吧：制作一张寂冷冥王星拼贴画
Get Crafty : Make an Icy Pluto Collage

制作一张拼贴画，展现出冥王星在冰雪皑皑的柯伊伯带中的样子。当然，你也可以再拼贴一个遥远的太阳。

天体可以用以下材料来表示：

- 彩纸或者纸板的碎屑
- 纸巾或者礼物包装纸
- 亮片

你需要：
- 一大张薄薄的纸板或者彩色美术纸（作背景）
- 剪刀
- 白胶
- 用来涂胶的刷子

下面是一张冥王星拼贴画的例子，你可以依此创作自己想象中的冥王星。想一想：

- 你要怎么展现冥王星表面的冰雪呢？
- 你要用什么来表现柯伊伯带中的冰封天体呢？

词汇表 Glossary

小行星 | asteroid
围绕太阳公转的大块岩石，有些小得像辆汽车，有些大得像座山。

小行星带 | asteroid belt
一个由小行星组成的巨大环状带，围绕着太阳旋转。

彗星 | comet
由冰、岩石和尘埃组成的天体，围绕太阳公转。

矮行星 | dwarf planet
围绕太阳运行的圆形或近圆形天体，比八大行星小得多。

飞越 | flyby
航天器近距离飞过行星、月球或者其他天体的行为。飞越过程中，航天器会靠近某一行星，以便对它进行仔细的研究，然后把信息传回地球。

冰川 | glacier
一块移动缓慢且巨大的冰体。

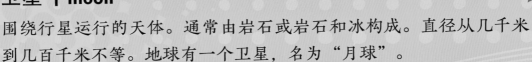

卫星 | moon

围绕行星运行的天体。通常由岩石或岩石和冰构成。直径从几千米到几百千米不等。地球有一个卫星，名为"月球"。

公转 | orbit

围绕另一个天体运行。

行星 | planet

围绕太阳公转的大型天体：一些行星，如地球，主要是由岩石组成的；其他的行星，如木星，主要是由气体和液体组成的。

探测器 | probe

不载人太空飞船。通常被送往行星或其他天体，用于拍摄照片并收集信息，由地球上的科学家操作控制。

自转 | rotate

物体自行旋转的运动。

太阳系 | solar system

太阳和围绕太阳公转的所有天体，如行星及其卫星、小行星和彗星。